Preface

Using a machine learning model called a recurrent neural network (RNN) and samples from a genuine literary work I generated text in the style and sentence structure of that literary work. The RNN trains on the source material and as it builds a new sentence it considers what words it has previously used, resulting in sentences and paragraphs that seem to make sense at a glance.

This work is part of the visual arts installation, "BitLit: Machine Poetry Corner" by Perth Machine Learning Group - Perth Fringe Festival, Australia, 2019. Will machines surpass the literary and creative skills of human artists?

Find out more at:
Web: www.pmlg.org
Twitter: @PerthMLGroup
LinkedIn: www.linkedin.com/company/perth-machine-learning-group
Meetup: www.meetup.com/Perth-Machine-Learning-Group

Principles of Machine Geology, 2019
A machine generated work.

Created by:	Hilary Goh, Perth Machine Learning Group
Using:	Recurrent Neural Network based on a GRU network Copyright 2018 The TensorFlow Authors
Source material:	Principles of Geology or, The Modern Changes of the Earth and its Inhabitants Considered as Illustrative of Geology By Charles Lyell

Principles of Machine Geology

⤌⟁ ⟁ ⟁⤍

Geology define - principal cause - seeds of a species - subsidence - cone of Vesuvius

CHAPTER I

Our knowledge of the Silurian isle or continent was appointed the equator, and nearly 600 miles west of the course of the climate of the more modern deposits. Yet even these have sometimes proved entirely destitute of mammifer.

Their speed must be always limited by the lavas of Etna. While moving on, its surface was in general consisted of the animate or inanimate world has been sometimes called "the age of reptiles," so great is the number of species nature.

It may nevertheless be advantageous to point to some existing anomalies in the geographical development to within very narrow limits, for alluded to, there appeared to me an obvious connection between the fauna now existing in any part of the globe. A greater change took place in its physical events, that the kind of lead us to infer the non-existence of creatures of which no monuments happen to sink where sedimentary rocks had been slowly deposited.

Next in chronological order above the Coal comes the allied Marge to assume that the earliest past have appealed, is reas, for example, as Chalk, Green Sand, Oolite, Red Marl, Coal, and the great of Stonesfield were supposed exclusively to belong constitutes the lowest grade in the class Mammalian boing on the earth, it is the intellectual and moral attributes of our race, rather than the physical geography of the globe, were referred to one and the same cities built and destroyed.[1]

The Argonautic expedition was despecies, and with a supposed progressive system, by which the order of nature, he would secondly, the progressive development of human power, or perhaps by some other new agent was with the inferior animals, but on his reason, by which it is said, and scratches on the ocean. It appears also that some of these ancient saurians approximated more nearly three to one in favor of our finding that such as the rampart as if still in the very act of falling.

In regard to the absence of birds and seas, and the grand mate of their formations were deposited when the physical geography of the globe, a greater change took place in its physical events cannot be traced in the same species for many years suggested that the *marsupial order* to which the age of the oldest stratified rock.

The strata of the Wealdenhange, or hills more than a thousand feet high. To explain the present state of science among men, and in aiding navigation of the same tertiary group, about 250 species of fossils and the passage from the more simple forms in each division of the animal or vaurient restion of a continent.

If we restrict ourselves to combinations of catania restriction of man is such a latitude parts of the world. At length, between the introduction of man into Asia, the supposed cradle of our race,

3 Principles of Machine Geology

and day be supposed to have a wider geographical range in the older strata, are not uncommon name and aquatic reptiles. There were iguanodons walking on the land and in the walls of Catania. These remains except those of fish, besides several kinds of cetacea.[2]

It may be asked, such a deviation from the antecedent course of physical events cannot. In France we meet with another fauna, both conchological and crocodiles, turtles, and *fossil fruits* implied a climate hot enough for the quadrumana, we had as yet no data formations were deposited when the physical geography of the northern hemisphere had been entirely and upper tertiary periods have been formed, as homogeneous in mineral composition throughout all time of the succession of sublunary events. Their accumulation elsewhere in regular strata, were all assumed to have taken place with an energy wholly unequalled in our times. These opinions, and the parallelism of the striate the ground all the houses in Nicolosi, a town situated near the lower margin of the woody region, about the human race, or the latitudes 52deg., as, for example, in Tierra del Fuego, as well as in the woody Regions.[3]

The deviation from the antecedent course of physical events cannot be traced in the same species for many years, suggested that the marsupial order to which this great current performed the former changes. But it was still supposed that denudation, or the power of the new agent was withheld, even for a brief period, a relapse would have remained engraven on the earth's surface has undergone.

Erratics. -- The next phenomenon ocean, which teem with life, may be said to have been the feeders of lateral cones, or, in other predaceous fish, such as were those of which we find the teeth preserved in some of the order of the ocean, to strip off superior strata, and lay the same kinds of rock have occasionally been reproduced at successive period.

The general absence of erraticsh in ancient rocks, that it has been evidently no part of the plan of Suffolk, a marine formation which has yielded three or four hundreds or thousands of feet above or below the rocks, as have the traces of the tenderest leaves of plants, and the soft integuments of many animals. The larger beasts of prey, in particular rocks of ancient date, it was almost unavoidable that this notion, when once emby those which frequent marshes, rivers, or the bords of a great variety of speculation, whether the entire suppression of one important class of vertebrata, such as compared to those of remoter eras, not because of a stream of lava.

A great philosopher has observed, that we can command naturalists of the Oolite, have been supposed to have a wider geographical range in the Carboniferous epoch, owing probably to the height of the two last-two miles, giving a velocity of *only twenty-two feet per hour*; and we are apt hastily to infer, as some writers have done, from a single extinct species.[4]

Cretaceous rock. -- Would have been traced back at least as far as those modern strata in which all the testacea, and of which a large proportion, although some of these islands are from 3000 to 4000 feet high, and one of them 75 miles long before the epoch of the Kimmeridge clay, an upper member of the Oolite. According to Professor Owen, it exhibited the continuance of the same species of fossils are more uniformly distributed throughout an indefinite lapse of ages, the epoch of the Kimmeridge another cause of deceptionnous state of things presents itself in the southern hemisphere. Even in the temperate zone, between the Lias and the Coal.[5]

The occurrence of these most ancient memorials of the

5 Principles of Machine Geology

mammifer was detected. At length, in the year 1839, there were formed with such instincts as are possessed by the lower animals.

If this be admitted, it would not followed each other in regular chronological order, the creation of each class being separated from the oldest Silurian to the newest Tertiary was as diversified as now. The shells, corals, and fossil plants, had been studied with an example of the ocean's bed"--"mankind had a beginning, since we can look back to the period when the surface of the earth, because "the strata of which our continents are composed were once a part of the plan of Stonesfield were supposed exclusively to belong constitutes the lowest grade in the class Mammaning by a single stroke, and not by a series of minor floods, the sinking of the barrier being gradual.[6]

A great philosopher has observedding to some existing anomalies in the geographical denied and course of nature. Thus we had no doubt of their veracity, might we not suspect them to be *unskilful naturalists?* or, in other predaceous fish, such as were those of which we find the teeth preserved in some of the surface of the earth.

The northern and southern hemisphere, as, for example, on the coasts of Canada and Gulf of St. Lawrence, and reach points of the western hemisphere farth the sea, or engulfed by the progressive development of human power, or perhaps by some other new ages in the older strata, are not uncommon aments and are marine deposits; nor can it be said that his remains are more perishable than those of other animals.

An intelligent being, after observing, of the human from all other species, considered merely as *physical agents*, with an energy wholly unequalled in our times. These opinions, and the having of the globe, a greater change took place in its physical events, that the kind of lead at the head of nature before the class of reptiles had been created. The confident assumption on geological grounds, so long as we are imperfectly acquainted with the order of superposition, that other strata, eruptions are frequent; whereas in the valley of Calanna, which is below that parallel, and in a fault, may be indistinguishable from one generation to another.[7]

But in North America still more ancient indications of the coal formation, we have at least ascertained that some of these ancient saurians approximated more nearly three to one in favor of our finding that such great problem considered in the preceding chapters, namely, whether the former changes of the earth at successive period the residence of human beings, was an eras to aqueous *causesna*. They seem clese of our race, and that distinct zoological and vegetable kingdoms; and the multitude of fossils dispersed through successive strata implied to the more perfect forms of animal organization and instinct.

To pretend that such a step, or high antiquity of man would have been inscribed in far more legible characters on the framework of the globe immediately before our species was called into existence, we find the evidence defective. There are ground all the houses in Nicolosi, about a state of animal life which might be called the converse of the extent would his belief in the regularity of the system be weakened? Would he ceased was detected so lately as September, 1852, by Mr. G. W. Dawson.[8]

And marine, like the seal, whale, and narwal; so in the early ages under consideration, there have been several *esuated* and sometimes wholly wanting, where birds, large land quadrupeds, and cetacea abound. We meet with another modified by degrees, and ages, one assemblage of animals and plants had disappeared

7   Principles of Machine Geology

after another again and again, and new treats of river-deltas and the dispersion of sediment by currents, and in the description of reefs of by griad alto appear to be as slight as was consistent with the conclusion.

That a small number only of mammalia inhabited European latitudes when once the agreement or disagreement in mineral character alone was relied on as yet another of the same order related to the phasmidae.

Mollusca and large animals from the opposite extreme, and denying that the earth might then have been some ground for speculating, on the former existence of a climate more severe than that now prevailing, in the same sense in which we believe that some of these belong to placoid fish, which occupy a high grade in the flood-season, a region capable of supporting a population of the animal kingdom, supposed to give way along the lines of least resistance, or where it was formerly rent asunder.

Silurian strata to view, should we be able to recognize, in man's entrance upon the earth's becoming at a time, would allow the sudden escape of vast floods of water into a hydrographical basin. Entire geologists are aware how large a proportion of all the events which may happen on this globe, would not enable a philosopher to speculate with the accompaniments of new and extraordinary circumstances, and those not of a *physical* but a marine.[9]
Carboniferous strata and lake any species of monkey, were first discovered near the Sutlej, in lat. 300 feet high. If, therefore, a series of subsidences should we not be skeptical as to the accuracy of their statements? and the same cities built and destroyed. There are regions at present, in the Indian and Preated long before the epoch of the Kimmeridge clay, is shown

by the Microlestes of the Trias before alluded to, soon perceive that no one of the fixed and constant laws of the animate or inanimate proof.

Contemplated from several other points of view mammalia depends; but nothing is more clear than that the causes which stamp so peculiar a characteristic of the animate world has been unceasing as the mammiferous, and the great development of another, such as the reptilian, implies a departure of the vinials of the same tertiary group, about 250 species of shells have been recognized, of which a still geologists that this ichthyic type was the more highly developed, because it took the reptiles do not predominate in number or size. The deposits formed at the more simple forms in each division of the animal kingdom preceded the cetacea, and discharged their functions in the animal economy.[10]

That mammalia had been detected until 1839, when the teeth of a leopard, a bear, a hog, and on the former existence of a climate more severe than that now prevailing in the western hemisphere farther species, considered merely as an efficient cause in the physical geography of the globe.

These opinions, and the articulation of the spinous processes with the vertebrae. Among living mammalia than do any of the reptiles now existing.[9] We are often misled, when we institute such comparisons, by our knowledge of the wide distinct series of subterranean movements. Time has been required, and a succession of events has taken place at two distinct periods. This standard of time, without which there are no means whatever of measuring the comparatively of modern origin. Is not the interference of the human from all other species, considered merely as an efficient cause in the physical geography of the globe.[11]

9 Principles of Machine Geology

No inhabitant of the land exposes which I have recommended in this work, that I am desirous of explaining the grounds for the belief entertained by many, that the intensity both of aqueous and of igneous forces, when the earth which as yet no bones of land quadrupeds have been obtained.

Regularly stratified sand and gravel, and others interfering with our agricultural and horticultural labors. We may have failed to discover a single shell, marine or freshwater, or a single coral or bone in ancient times, is founded the creation of the more simple forms in each division of the animal kingdom preceded the cetacea, and discharged their functions in the animal economy.

Nor was the more highly developed, because it took the deposition and denudation are parts of the same grave.[12] But even if the more solid walls. A gentleman of Catania, named Pappalardo, desiring to secure that he had at once become fertile upon the landing of the same manner, and in the destroying causes in the Val del Bove; for there is nothing which afford a very imperfect parallel to the state of the animal kingdom, supposed to have been inscribed in far more legible characters on the framework of the earth, and the laws which govern its animate productions, as in 1832, when they ran in all directions from the centre of the volcano.

It has lately been ascertained that glaciers give rise to these effects when we speculate on the state of America in the induction of a common standard of time, without which there are no means whatever of measuring the comparative.

Transportation of blocks by ice, the state of things to have gone on according to the order now observed in regions unoccupied by man. If such a fact tends to destroy all our confidence in the uniformity of the order of nature, he would the first introduction of the fragmentary matter. Nor should we forget that the last term of one and the same series of progressive developments?

In reply to this question in the scale of beings now dwell in various quarters of the globe where marine species lived within the times of history, been buried under volcanic ejections, submerged beneath the sea, or engulfed by Mr. Owen to belong to pterodactyls.[13] But in North America still more ancient indications of the animal and vegetable world, from the simplest to the most perfect forms, rests on a very imperfect parallel to the state of the animal kingdom, supposed to have a wider geographical range the accompaniments of new and extraordinary circumstances, and those not of a physical but a moral nature.

The deviation permitted transportation of blocks by ice, some of the older lavas of Etna; for the light emitted from the great rent of S. Lio appears to in the flood-season, a region capable of supporting a population of many millions might be suddenly at particular periods; and, secondly, that the contemplated from several other periods? Now these objections would be unanswerable, if adduced against one who was contending to identify, in age, all the red sandstones and marls in question, has at length been sufficient time since their organization, by which individual peculiarities are transmissible from one generation to another.[14]

The occurrence of these most ancient memorials of the mammifer, have fancied alterations still more remarkable in the economy of nature to have attended the first adventurers on the shores of the New World. In that interval, we imagine that the operations of water were on a different and grander scale in ancient times, is founded on single specimens, while a still greater number are founded on single specimen.

11 Principles of Machine Geology

Suppose our mariners were to report than any part of Great Britain. The general absence of erratics in the older strata, are not uncommanian relations which may be supposed to have attended the first extent of the same convince as the mammiferous, and the great development of another, such as the reptilian. Implies a departure from the antecedent course of physical events cannot be traced in the same species of fossils are more uniformly distributed throughout all time of the succession of sublunary events--if, for example, now seen stretching for the order of precedence.[15]

In the course of caverns in the position of land and sea, accompanied by great fluctuations of climate. To these questions it may be answered, that the agency of man did not constitute an anomalous deviation from the same kind of sediment, is now actually distributed by the occasional present counts of the animate world, has been unceasing the development of certain instincts, or by availing ourselves of that mysterious law of their form would have.

Can we expect for a moment, and at one strata in the animal provinces, as they are called, which form so striking a few bones of aquatic of the same class of reptiles may of progressive development.[16]

Or to the concepenent of the fossil remains formerly referred to the amposite extreme, and denying that the earth might then have been said, that the intensity both of aqueous and of igneous forces, whenever the power of the habitations of livers of the animal world, is to strain analogy between the former conditions of the animate world has been unceasing the development of certain instincts. By availing ourselves of that mysterious law of their form would have been traced back at least as was called into existence, we have at least ascertained that some vessel had been wrecked at land and the parts of Europe, has become very our residence.

If the barren soil around Sydney had at once become fertile upon the landing of the same class. The remains of fish are as yet confined to the land, the class of reptile.[17] Igneous force snow, such a fact tends to destroy all our confidence in the uniformity of the course of animal and vegetable existence required for such a climate more severe than that now prevailing in the western hemisphere which had as yet appeared upon our planet.

The fact, it was said, confirmed the theory which has the learn from Dolomieu that the stream moved during part of its course at the rate of 1500 feet and always recollect that a climate like that now experienced at the equator, coexisting with the uniformity of the course of animal and vegetable existence required for such a climate more severe than that now prevailing in the western hemisphere which had as yet appeared upon our planet.

It was said, confirmed the theory which has learned from Dolomieu that the stream moved during part of its course at the rate of 1500 feet an hour, always recollect that a climate like that now experienced at the equator, coexisting with the uniformity of the course of animal.

Our contemporaries on the earth, and we may, therefore, ask whether his creation can be considered in the light of a disturbance or deviation from the system, than the oolite of Stonesfield were supposed exclusively to belong constitutes the lowest grade in the class Mammalia depends.[18]

But nothing is more clear than that the causes which stamp so peculiar a characteristics of sedimentary rocks, their organic

13   Principles of Machine Geology

remains, many naturalists of a land quadrupeds, and composition, were of contemporaneous date, it was thought sufficient. It was in vain to the sea and land ever cease while the more simple forms in each division of the animal kingdom preceded erratics into their present situations.

CHAPTER II
The causes which may be supposed to given the oolitic texture was declared to be rather an exception than otherwise to the general rule in rocks of a given age than was really the case. There is still another cause of deception, disposition as those of horses which were buried in the same grave.[19]

Scanty as is the information hitherto from the present, we ought not to look for any anomalous results, unless where man has interfered, or unless clear indications of the land or streams, was by the occasional present causes are in existence.

Terrestrial species, therefore, might be older than the continents and islands as expose Silurian strata to view, should he considered as such a deviation from the analogy of those affected by other animals than is usually supposed to have prevailed during the secondary periods, when a high temperature pervaded European latitudes, was a state of things to which there is now no counterposed in the first edition of this work, January, 1830.[20]

In opposition to the theory of progressive development, another and a far more difficult one may arise our species, and causing the fauna and flora to pass from an embryonic to a more point out in the first edition of this work,[21] in which I stated that the bones of quadrupeds have been obtained.

In the present imperfect state? If we concede like the

hippopotamusy other classes of organic remains demonstrate the fact that the earth might then have been divided from the present, we ought not to look for any anomalous results, unless where man has interfered, or unless clear indications of the land or streams, was by the occasional present causes are in existence.[22]

The hippopotamus to the same causes, and the conveyance of solid matter to a particular region can only keep pace with the newer Pleiocene mammalia in Europe, they belong decidedly to many different epochs? Or what security have we that they may not arise hereafter?

And if such be surprised if the only accessible strata should be limited to deposits formed far from land, because we have not yet found a single Silurian helix, insect, bird, terrestrial reptile or mammifer, have fancied alterations still more remarkable in the economy of nature to have attended the first of these cases, we may take Lake Superior, which is more than 400 geographical miles in the rocks, as have the traces of the tenderest leaves of plants, and the soft integuments of many animals.[23]

The larger beasts of prey, in particular attention. Analogy led every naturalist to assume, that each full-grown individual of the animal or ving mammalia depends; but nothing is more clear than that the causes which stamp so peculiar a character.

There were iguanodons walkinghous type, in so low a member of the oolitic series, while no other representatives of the same class. The remains of fish are as yet confined to the upper particular period the residence of human beings, was an era in the moral, not in the physical world as repeated at distant intervals, which have been thought by some to have come from their original position.[24] For between them and the parent

15 Principles of Machine Geology

rocks we now find, not unfrequently, deer, and ox, while species which he shares in common with the inferior animals, but on his reason, by which he is distinguished geologist, M. Elie de Beaumont. In several essays on this subject, the last term of one and the same series of progressive developments?

In reply to this question in the Red Crag.[25] Of a still newer were terrestrial, winged, and aquatic reptiles. There were iguanodons walking on the land, pterodactyls.[26] But in North America still more ancient indications of the animate world has been unceasing to remain. Professor Forbes has confined to the upper parts of the southern hemisphere, where the indigenous land quadrupeds are comparatively few, and of mammalia without reptiles were not created until after the close of the carboniferous epoch.[27]

In the course of that years and extreme improbability the chance of our hitting on those minute points of space where caves may of England, occurs as a member of the Carboniferous group, and in the United States near the Falls of Etna. Here ancient indications of the animate world has been unceasing at present derived entirely from rocks older than the Eocene has been frequently of many distinct series of subterranean movements.[28] Time has been required, and a succession of a physical but a peculiar fossils may modify or entirely overthrow all our presence as before any isle or continent was appointed the crocodiles, turtles, and fossil.

Comprising the Permiank, the Upper New Red Sandstone of English geologists, belong, nevertheless, to the Eocene epoch, such as the remains of the Palaeotherium, Anoplotherium, and other extinct quadrumana of the same date.

16   Principles of Machine Geology

Since those remarks were first student was taught to understand them in no other than a chronological sense; so that the Chalk might be geological evidence, of the first time at an era subsequent to the carboniferous.[29]

Scanty as is the information hitherto met with in tertiary deposits were chiefly those which frequent marshes, rivers, or the bords, traversed by numerous open fissures. Part of Catania destroyed the continuance of the same species for many generations. At length the idea that species themselves to a disturbed imagination during the visions of the night.[30]

Until lately it was supposed to be a marine letter, a combination of existing causes may have conveyed erratics into their present situations.

The causes which may be supposed to give way along the lines of least resist from the earliest periods seems they are mater; and second, the upward movement of the bed of the sea, converting it gradually into land. With though a minority of the whole number, are recent, besides many corals, echini, foraminifera, and fish, but no signs of previous trituration, so that neither its insects winged their way through the ancient forests.[31]

In the ironstone of Coalbrook Dale, two species of cetacea of a new genus, which is comparatives, those sandy tracts had begun to yield spontaneously an annual supply of grain, we might then, inder the small area of the earth's surface hitherto explored geologically, and the new discoveries brought to light by geology.[32] Thus in the arctic regions, at present, reptiles are small, and that the old red sandstone, or Devonian rocks, contained no vertebrate remains except those of fish, besides several kinds of chelonian and saurian residence in advance, no elevation in the scale of beings on the earth, it is the intellectual

17 Principles of Machine Geology

and moral attributes of our race, rather from the earliest periods and consequently depressed below the Black Sea and Mediterranean. This area includes the site of the Produced a reptile, called by him Apateon, related to the salamanders; and in 1847 three species of another genius of Shakspeare.[33]

So daring a disregard of probability and violation of analogy would have been once those effects usually called diluvial; for the difference of level of 600 feet between Lake Superior, which is more than 400 geographical miles in the rocks, as have the traces of the tenderest leaves of plants, and the soft integuments of many animals. The larger beasts of prey, in particular rocks of ancient date, it was almost unavoidable that this notion, when once embling a corydalis, together with another of the same order related to the phasmidae. As an example of the insectivorous arachnidae, I may mention the scorpion of the Bohemian coal, figured by Count Suffolk, in lat. 52deg,[34] in the London clay, the fossils of which, such as crocodiles, turtles, and fossil.

In modern times, no rivers have excavated of the animal and vegetable world, from the simplest to the most perfect forms, rests on a very in the heavens, the same identical phenomena recurred again and again in a perpetual vicissitude. The ordirar and mounded solely on negative evidence.[35]

So important an exception to a general rule may be perfectly considered chronologically, has proved that the terrestrial vegetation of the Carboniferous epoch, owing probably to have been inclosed in the stalagmite of caverns in the older Pleiocene, or in the Miocene or Eocene deposits mark the era of the first creation of quadrumana.[36] It would be reasonable to draw such inferences with respect to the future only by obeying her laws; and this principle is true even in regard to the astonishing

18 Principles of Machine Geology

changes which I have recommended in this work, that I am desirous of explaining the grounds for the belief entertained by many, that the intensity both of aqueous and of igneous forces, the stand near the rocks, as have the traces of the tenderest leave.

Currents, for example, tides, to be surprised if the only accessible strata should be limited to deposits formed far from land, because we may inquire, for example, whether there are any great changes in the land quadrupeds inhabiting Europe, probably not less than five complete revolution is effected in a brief period than the first edition of this work, January, 1830, in opposition to the theory of progressive der the vertical fissures now filled with rock to have been the feeders of lateral cones, or, in other periods?

Now these objections would be unanswerable, if adduced against one who was contending to identify, in age, all the red sandstones and marls in question, has at length been sufficient time since their organization to the type of living mammalia than do any of the reptiles now existing.[37]

It is probable from these and many other considerations, that as we enlarge our knowledge of the Silurian fauna is at present derived entirely from rocks of marine origin, no from the reduction of rocky masses. That we should find, therefore,ht of an orifice not far distant from Monti Rossi, which at that time opened and poured out a lava current of 1669, before alluded to, soon the waters of the ocean may be raised by these convulsions, and then break in terrific waves upon of all these groups--their identity in mineral composition was thought sufficient.[38]

It was in vain to the sea was scanty at that era, but because in general the preservation of any relics of the animals or plants in past time coincides with the age of the oldest stratified rock in

19   Principles of Machine Geology

which the geologist was in the act of accumulating, and if the numerous causes of subsequent disintegration should not other changes which I have recommended in this work, that I am desirous of explaining the grounds of the Oolitic period.

It often became necessary to affirm that no particle of carbonaceous matter could be detected in districts where the true Coal series abounded.[39] In spite of every precaution of species now living.

Faults. -- The same reasoning is application leopardy, the first living species proper to the ocean has been discovered of the gradual passage of the earth from the oldest Silurian to the newest Tertiary was as diversified as now.[40] The shells, corals, and fossil plants, had been studied with an energy wholly unequalled in our times. These opinions, and then break in terrific waves upon the economy of nature in a few months; and, consequently, as their mechanical and derivative origin was already admitted, but it was still supposed that denudation, or the power of the human race is extremely modern, even when compared to the larger number of species now living.[41] The borden of part of the unfathomable ocean into a shoal rather leap, can be part of a regular series of changes in the animal world, is to strain analogy between the fauna now existing in any part of the globe, a greater change took place in its physical movements has yet gone, we have every reason to suppose, the physical movement must be guided by the same rules of induction as when we speculate on the state of America in the inhabitants of that town.

Being along with it sand, gravel, and stony fragments, some of them hundreds of tons in we arrive at the coralline crag of Suffolk, a marine formation which has yielded three or four hund on the other hand, the sea now prevails permanently over large districts once in England faults, in which the vertical displacement is between 600 and 3000 feet, and the horizations of its species by generation; and thus the first approach was made to the conception of a common standard of time.[42]

Without which there are no means whatever of measuring the comparting to identify, in age, all the red sandstones and marls in question, has at length been sufficient time since their organic remains, many naturalists of the ocean's bed"--"mankind had a beginning, since we can look back to the period when the surface of the earth, because "the strata of which our continents are composed".

The lava current of 1669 before alluded to, soon perceive that no one of the fixed and constant laws of the animate or inanimate world existed, affords ground for concluding that the experience during thousands of ages by inferior animals.[43] Why should not other changes which has yee globe, were referred to one and the same series of progressive development, another and a far more difficult one may arise out of the observed fact that sets of strata of different ages. One assemblage of animals and plants had disappeared after another again and again, and new tribes had started into life to replace them. [44]

Denudation. --In addition to the proofs derived from one century to another to determine the growth of certain tribes of plants and animals can be proved to have existed, there has been a continual changes which I have recommended in this work, that I am desirous of explaining the grounds of many distinct series of subterranean movements. Time has been required, and a succession of events has taken place at.

The End.

Foot notes:

[1] Prichard, vol. i. p. 277.

[2] Darwin's Journal, p. 153., 2d ed. p. 133.

[3] Ibid. vol. i. p. 277 p. 177.

[4] Phil. Trans. 1832, p. 240.

[5] Prof. J. D. Forbes, Edin. Journ. of Sci. No. xix. p. 130, Jan. 1829.

[6] Darwin, Structure and Distrib. of Coral reefs, &c., London, 1832. It was referred to by Sir J. M. South America, vol. i. ch. 2, a colored geological map and section of the Niagara des Assain; and the land between the 30deg N. lat. and the pole in former fluctuations of climate at successive geological periods, agrees in every essentioned by the artist to represent the conclusion, that at a moderate distance from the shore, the conjectured, and that a hore recent was scarcely ever been outdone essentis figuram imitari. p. 18.

[7] Scrope on Volcanoes, p. 109.

[8] Ann. de Chim. et de Phys. tom. iii. p. 287.

[9] Prichard's Phys. Hist. of Mankind, vol. i. p. 275.

[10] Hist. Mundi, lib. ii. c. 107.

[11] Buckland, Reliquiae Diluvianae, p. 25.

[12] See above, p. 400.

[13] See remarks by the late Sir Alexander

   Burnes for the accompanying sketch (fig. 72) of the fort of a

   gradual transition has been the probable depth of the sea

   from some of the phenomena of Earthquakes, Phil. Trans.

   1835, p. 329.

[14] See De Beaumont, Mem. pour servir, &c. tom. iv. p. 187.

[15] De la Beche, Geolog. Manual, p. 82.

[16] Von Hoff, vol. i. p. 366.

[17] Darwin's Journal, p. 153., 2d ed. p. 133.

[18] Fleming, Ed. New Phil. Journ. No. xlviii. p. 135.

[19] Mr. Forbes, Account of Mount Vesuvius, Edin. Journ. of Sci. No. xix.

   p. 53.

[20] See Palmer on Shingle Beaches, Phil. Trans. 1836, p. 35.

[21] Von Buch, Descrip. des Iles Canar. p. 450, who cites Erman

   at the sea figure in plate 40 of Sir W. J. Hooker, in his

   conveysion, high the sun's centre, the heat received in

   describing any part of its orbit, is proportional to the angle of

   the editor of the month of the sea from some of the

   phenomena of a great change produced in the second volume

   of "Observations made in the Indies by the earthquake

[22] Mediterranean. Prichard, Phys. Hist. of Mankind, vol. i. p. 34.

23 Principles of Machine Geology

[23] See Lyell's Manual of Elementary Geology, ch. 20.

> exceedingly diminished at the depth of twelve or fifteen feet. He are told, were "a musaeo subducta, annuente vice cancellarks, by Mr. Toomer, who accompanied Captain Kelowa, chap. iv.

[24] Phil. Trans. 1836, p. 155.

> Essay the meatural history that it deserves commemoration, and it is with no small interest that we learn, and of the other dotted.

[25] Quart. Journ. of Agricult., No. ix p. 433.

> For an account of the more modern changes of the tertiary found to yield uniform and constant resume the esture, p. 409.

[26] Ibid. p. 2.

[27] See Manual of Geol. ch. xi.

[28] Agassiz, Jam. Ed. New Phil. Journ. No. lii. p. 28.

[29] Reduced, by permission, from a figure in plate 40 of Sir Honseivers, by Capt. Henry M. Denham, R. N., who reached bottom at 7706 fathoms (46,236) Dr. Geol. of East Norfolk, p. 10.

[30] Essay on the Habitable Earth, Amoen. Acad., vol. ii. p. 220.

[31] Asiatic Journal, vol. i. p. 41.

[32] Hort. Trans. vol. ii. p. 220.

[33] This is shown by projecting a map on the horizon of Bonito the Principles, 1846, p. 12.

[34] See Dr. Buckland's description of the Earth, &c. P. 229. Some of the principal fact that the submerged for the account of Flands, and the Colifer curtains and very some specimens of supposed palm would not have happened to two fools."

[35] Book iii. ch. 50.

[36] Essay on the Habitable Earth, Amoen. Acad., p. 5.

[37] Agassiz, Jam. Ed. Voyage of the Beagleor Soc. vol. i. p. 24.

[38] Nat. Hist. lib. iii. c. 2.

[39] Geog. des Plantes. Diet. des Sci.

[40] See Catalogue of British Association, 1846, p. 187.

[41] Nov. Com. Petrop. vol. xvii. p. 45.

[42] Proceedings Roy. Irish Acad. 1846, p. 26.

[43] Journal of a Residence in Iceland, p. 276.

[44] Trans. of the Academy of Sciences, January, 1837, on various subjects, relating to fossil remains and the effects of the height assigned to it by Recupero, refused to acquiesce in landing on it. The seven rival names are Nerita, Ferdinanda, Hotham, Graham Islands, p. 115.

www.ingramcontent.com/pod-product-compliance
Lightning Source LLC
Chambersburg PA
CBHW031512210526
45463CB00008B/3207